CONTENTS

EUROPE

1. Which is the most densely populated country in Europe?
a. Malta   b. Vatican City   c. Monaco

2. Other than The Vatican City, which European country is surrounded by Italy?

3. Which European country is home to Transylvania?

4. Name three European countries who use the Euro as their official currency despite not being members of the European Union.

5. Which city hosted the 2012 Olympic Games?

6. Which Nordic country has the oldest monarchy in Europe?

7. Which country has the longest coastline in Europe and the second longest in the world?
a. Greece   b. Greenland   c. Norway   d. Italy

8. Victor Hugo, poet, and writer of Les Miserables, has a street named after him in every city of which country?

9. By population, what is the largest city in Scotland?

10. Name the only island county of England (not the UK).

11. The Azores are an autonomous region of which European nation?

12. Which European country has the most volcanoes?
a. Russia   b. Italy   c. Romania   d. Norway

13. Which European country is divided into 26 cantons?
a. Austria   b. Albania   c. Switzerland

14. Name Europe's highest peak.
a. Mont Blanc   b. Mount Elbrus   c. Ushba

15. What was the official currency of Italy before it adopted the Euro in 1999?

16. Name the European country which is landlocked by other landlocked countries (and therefore doubly landlocked).

17. Where would you find the city of Valletta?

18. Name the largest island in the Mediterranean Sea.

19. In which country would you find the southernmost part of mainland Europe?

20. Name the country which is bordered by Russia, Poland, Ukraine, Latvia, and Lithuania.

21. In which country would you find the fishing village of Reine?
a. Denmark   b. Sweden   c. Norway

22. Standing at 3,798 metres above sea level, Grossglockner is the highest mountain of which country?
a. Germany   b. Austria   c. Switzerland

23. In which city is the La Sagrada Família located?
a. Madrid   b. Barcelona   c. Lisbon   d. Florence

24. Amsterdam is renowned for its cycling, but how many kilometres of cycle paths does the city approximately have?
a. 9000   b. 21,000   c. 35,000   d. 53,000

25. In which country would you find the Spanish Steps?
a. Spain   b. Portugal   c. Italy   d. Hungary

26. The Vatican City has the lowest high point of all European countries. How high is its high point?
a. 46 metres   b. 75 metres   c. 97 metres

27. As of September 2021, which of the following countries was the most recent to join the EU?
a. Romania   b. Croatia   c. Bulgaria

28. The United Kingdom consists of England, Wales, Scotland and Northern Ireland. Which countries does Great Britain consist of?

29. Which French city has the nickname Pink City?

30. Which landlocked country borders Ukraine and Romania?

31. What is the largest lake in Europe?
a. Lake Ladoga   b. Lake Garda   c. Lake Como

32. In which country is Lake Annecy located?

33. Where can you find the Saxon Switzerland National Park?
a. Switzerland   b. Austria   c. Hungary   d. Germany

34. According to the United Nations, how many countries are there in Europe?
a. 28   b. 35   c. 44   d. 49

35. Which European country most recently gained independence?
a. Montenegro   b. Serbia   c. Kosovo   d. Ukraine

36. In which year did the European Union start?
a. 1990   b. 1991   c. 1992   d. 1993

37. When did the UK hold a public vote, where the country voted to leave the EU, also known as BREXIT?
a. 2015   b. 2016   c. 2017   d. 2018

38. How many stars appear on the European Union flag?
a. 5   b. 8   c. 12   d. 20

39. As of September 2021, how many people live in Europe?
a. 587,342,983   b. 748,186,266   c. 912,523,994

40. What is the most westerly country in Europe?

AFRICA

1. Which African country has the most pyramids? Hint: It's not Egypt.

2. Name the African country which is landlocked by South Africa.

3. What is the most populated country in Africa?
a. Kenya   b. Ethiopia   c. Egypt   d. Nigeria

4. Which of the Seven Summits can be found in Africa?

5. What is the easternmost country in mainland Africa?
a. Somalia   b. Mozambique   c. Tanzania   d. Kenya

6. Name the African country with the same name as its capital.

7. What is the northernmost African country?

8. Name the African nation which gained independence from Ethiopia in 1991.
a. Sudan   b. Uganda   c. Eritrea   d. Mali

9. In which country would you find the city of Kampala?
a. Namibia   b. Guinea   c. Uganda   d. Gabon

10. What is the largest country by area?

11. What is the most populous city in Africa?
a. Lagos   b. Nairobi   c. Kinshasa

12. Name Africa's famous desert which is the largest in the world and is the same size as continental USA.

13. Which African nation has more Portuguese speakers (by around 8 million people) than Portugal itself?

14. Name Africa's smallest country.
a. Liberia   b. Sierra Leone   c. Seychelles

15. Name Africa's smallest mainland country.
a. Benin   b. Gambia   c. Eswatini

16. Name the three African nations with their own monarchies.

17. In which country would you find the city of Dodoma?
a. South Sudan   b. Gabon   c. Tanzania

18. Name the country which is bordered by South Sudan, Ethiopia, Somalia, Tanzania and Uganda.

19. Which word can be found in the titles of three African countries?

20. This country was once called Dahomey. Which country is it?

21. What is the westernmost country in Africa?

22. Which country is the main home of the Shona people?

23. In which African country did Germany commit the first genocide of the 20th century in?
a. Kenya   b. Tanzania   c. Namibia   d. Egypt

24. Which country is Serengeti National Park in?
a. Tanzania   b. Niger   c. Mali   d. Ghana

25. Which country was formerly known as Rhodesia?

26. Where would you find the city of Timbuktu?
a. Madagascar   b. Mali   c. Guinea

27. By population, what is the largest city in Morocco?

28. Cairo is the largest city in Egypt by population, but what is the second largest?

29. By population, what is the largest city in South Africa?

30. Which of the following does not border the Sahara Desert?
a. Black Sea   b. Atlantic Ocean   c. Pacific Ocean   d. Bering Sea

31. Which river in Africa flows into the Indian Ocean?
a. Kowie River   b. Mania River
c. Zambezi River   d. Luangwa River

32. Which of these ethnic groups of people can be found in Burkina Faso?
a. Mossi   b. Ewe   c. Yoruba

33. Which canal connects the Mediterranean and the Red seas?

34. What is the dominant religion in Africa?

35. Which of the following languages is spoken on the East coast of Africa?
a. Malagasy   b. Swahili   c. Songhay

36. Despite being a very large continent Africa has a relatively short coastline, but where does its coastline rank compared to the other continents?
a. 4th   b. 5th   c. 6th   d. 7th

37. Africa is the most centrally located continent in the world. It lies on the equator (0 degrees latitude), and on 0 degrees longitude. What is the term for 0 degree's longitude?

38. True or false: Africa is home to the hottest country in the world based on average temperature.

39. How many countries does Africa consist of?
a. 38   b. 46   c. 54   d. 62

40. Approximately how many different ethnic groups are in Africa?
a. 2000   b. 3000   c. 4500   d. 7000

NORTH AMERICA

1. Which U.S state has the most active volcanoes?

2. In the U.S, what is the only state with just one syllable?

3. Name the National Park which is mostly in the state of Wyoming, but also stretches into Montana and Idaho.

4. Name the second largest US state.

5. What is the largest of all US overseas territories?

6. Complete the country name: Antigua and _____.

7. In which Canadian province would you find Toronto, the most populous city in Canada?

8. Name the Valley which is the hottest, lowest, and driest area in North America.

9. Name the Caribbean Island which was named after the Latin name for 'Sunday', as this was the day Columbus discovered it.
a. Antigua   b. Barbuda   c. Dominica

10. Name the most eastern Caribbean Island. It is also the only Caribbean Island to be entirely in the North Atlantic ocean.
a. Cozumel   b. Barbados   c. Cayo Coco

11. What is the most populous country in the Caribbean? It is also the only Caribbean country to have a railway.

12. What is the only Central American country to not have a coastline on the Caribbean Sea?
a. El Salvador   b. Guatemala   c. Panama

13. Which Central American country is the only one to have English as its official language?
a. Costa Rica   b. Nicaragua   c. Belize

14. Name the largest of Canada's three territories.

15. Name the only Great Lake to entirely be in the United States.

16. What is the least populated state in the U.S?
a. Wyoming   b. Vermont   c. Alaska

17. What does the 'A' stand for in 'ABC Islands'?

18. Name Central America's largest country.

19. Complete the Canadian province: Newfoundland and _____.

20. Which three oceans border North America?

21. How many colonies were there originally in the USA?
a. 12   b. 13   c. 14   d. 15

22. How many time zones are there in mainland North America?
a. 6   b. 8   c. 10   d. 12

23. Which country has the most land mass in North America?

24. Which famous off road motorsport race happens in the Baja Peninsular?

25. Which group comprises of the largest indigenous people in Mexico?
a. Nahuas   b. Mayans   c. Zapotec

26. How many states does Mexico have?
a. 13   b. 19   c. 26   d. 31

27. What is the national animal of Mexico?
a. Cheetah   b. Mountain lion   c. Jaguar

28. There are five US states that touch the Pacific Ocean. These are Washington, Oregon, California, Alaska and which other state?

29. In which state is Mount Rushmore located?
a. North Dakota   b. California   c. South Dakota   d. Oregon

30. Which state is the only US state to have a rain forest?

31. Who visited Newfoundland in 1000BC, and were the first Europeans to lay eyes upon the New World?
a. Spain   b. England   c. France   d. Sweden

32. In which state is the eighth wonder of the world, Niagara Falls, located?

33. How many national parks is Alberta home to?
a. 3   b. 4   c. 5   d. 6

34. What is the capital of British Columbia?

35. How many countries does North America have?
a. 12   b. 17   c. 23   d. 29

36. There are two North American countries that both begin with the letter H. Name both.

37. Approximately how many islands are there in the Caribbean?
a. 100   b. 400   c. 700   d. 1300

38. True or false: Canada has a larger population than Australia.

39. Which country is the southernmost point in North America?

40. Which country has the fourth largest population in North America?
a. Canada   b. Nicaragua   c. Cuba

ASIA

1. Which country occupies 80% of the Arabian Peninsula?

2. Which is the largest island in Southeast Asia, which is home to three countries?

3. Which city is claimed by both Israel and Palestine?

4. Which Asian country is larger than France, Spain and Germany combined, but has a population of just over two million people.

5. This is the largest archipelagic country in the world, with over 17,000 islands.

6. Which Asian country is doubly landlocked (locked by other landlocked countries)?
a. Turkmenistan   b. Uzbekistan   c. Kazakhstan   d. Bhutan

7. What is the name of the desert in East Asia that covers North-eastern China and Southern Mongolia?

8. Name the administrative region which is the gambling capital of the world and is sometimes known as the 'Monte Carlo of the East'. Gambling tourism makes up 50% of this autonomous region's economy.

9. What is the currency used in Thailand?

10. Which country has the world's second largest Muslim population?
a. India   b. Afghanistan   c. Iran   d. Pakistan

11. What is the smallest country in Asia?
a. Bahrain   b. Singapore   c. Maldives

12. Which country is the smallest country to occupy Borneo?

13. Which is the smallest of the 'stan' countries?

14. Which is the only country in Southeast Asia not to be colonised by Europeans?

15. Which is the most southern Middle Eastern country, which borders Oman and Saudi Arabia?

16. Name the Asian country which is bordered by Georgia, Russia, Armenia, Iran and Turkey.

17. In which Asian country would find the city of Dhaka?

18. Which island is the most popular tourist destination in Indonesia?

19. What is India's richest yet smallest state by area, which borders the Arabian sea?

20. Which of these is an island country which lies on the Tropic of Cancer?
a. Sri Lanka   b. Philippines   c. Taiwan

21. How many countries make up Asia?
a. 48   b. 54   c. 61   d. 72

22. What percentage of the world's population live in Asia?
a. 43.5%   b. 50.2%   c. 59.8%   d. 66.8%

23. Which of the following does the Mekong River not flow through?
a. Vietnam   b. Laos   c. Mongolia

24. Which of the following cities lies on the Saigon River?
a. Ho Chi Minh City   b. Hong Kong   c. Beijing

25. Which is the only Southeast Asian country that is landlocked?
a. Malaysia   b. Vietnam   c. Laos   d. Thailand

26. Which city hosted the 1964 Olympic Games?
a. Beijing   b. Bangkok   c. Tokyo   d. Hanoi

27. The Ringgit is the currency for which Asian country?
a. Singapore   b. Cambodia   c. Malaysia

28. In which country is Dhivehi the official language?
a. Philippines   b. Oman   c. Maldives

29. What is the lowest body of water on earth that is found in Asia?
a. Caspian Sea   b. Dead Sea   c. Lake Zaysan

30. What is the name given to the art of folding paper to create three dimensional objects?

31. Which mountain in Indonesia is the highest island mountain in the world?
a. Mount Pinatubo   b. Jaya Peak
c. Mount Indomia   d. Mount Rinjani

32. Which is the holiest city in Islam?

33. Approximately how many years did it take to build the Great Wall of China?
a. 1 year   b. 5 years   c. 20 years   d. 100 years

34. Sirimavo Bandaranaike became the first female Prime Minister of which country in 1960?

35. What is the longest river in Southeast Asia?

36. How often are the Asian Games held?

37. In kilometres, how far is Sri Lanka from India?
a. 3   b. 29   c. 165   d. 712

38. Which of the following countries does not border Thailand?
a. Cambodia   b. China   c. Laos   d. Myanmar

39. What language is the most spoken in Bangladesh?

40. How many islands make up the Maldives?
a. 375   b. 634   c. 910   d. 1190

SOUTH AMERICA

1. What is the largest and most populated country in South America?

2. There are two South American countries that are landlocked. Can you name both?

3. Which is the only country in South America with an Atlantic and a Pacific coast?

4. Name the longest mountain range in the world which is found in South America.

5. Name the highest peak in the Americas which is found in Argentina.

6. Which country was previously known as New Granada?

7. Easter Island is a territory of which South American country?

8. In which country would you find the source of the Amazon River?
a. Ecuador   b. Peru   c. Bolivia   d. Chile

9. What is Argentina's desert called?

10. Name Venezuela's famous waterfall which is the highest in the world.

11. Which South American country uses the US Dollar as their official currency?
a. Paraguay   b. Colombia   c. Ecuador

12. Name the desert which is found in Chile and is considered the driest place in the world.

13. Which European country has an overseas region found on the northeast coast of South America?

14. What is the official language of Brazil?

15. Which Islands are located off the coast of Argentina?

16. Name the most populous city in South America.

17. The capital of this country is Paramaribo.

18. What is the capital of Guyana?

19. What is the capital of Colombia?

20. What is the largest lake in South America?
a. Mar Chiquita   b. Lake Titicaca
c. Lago Argentino   d. Viedma Lake

21. How many countries share a land border with Brazil?
a. 5 b. 7 c. 9 d. 11

22. How many countries does South America consist of?
a. 10 b. 12 c. 14 d. 16

23. Easter Island is part of which country?
a. Chile b. Ecuador c. Paraguay

24. Which country is home to the pre-Colombian Nazca lines?
a. Uruguay b. Peru c. Argentina

25. To which country do the Galapagos islands belong?

26. South America is home to the world's largest salt flat, which is called Salar de Uyuni. In which country can this be found?
a. Chile   b. Brazil   c. Peru   d. Bolivia

27. What is the official currency of Ecuador?

28. Where would you be in South America if you were spending euros?
a. Falkland Islands   b. French Guiana
c. South Georgia   d. Sandwich Islands

29. In which country would you find the Andean city of Cusco, that has an elevation of 3400m?

30. Which of the following cities has never been the capital of Brazil?
a. Brasilia   b. Rio de Janeiro   c. Salvador   d. Sao Paulo

31. True or false: Angel Falls is the world's widest waterfall.

32. What is an arepa? It is popular in Venezuela and Colombia.
a. A local fish   b. An omelette
c. A corn pancake   d. A local white bread

33. Which country in South America is the greatest producer of quality emeralds in the world?
a. Colombia   b. Uruguay   c. Venezuela

34. What is the name of the 18th century rebel leader in Bolivia, who was the last indigenous leader of the Incas. This person also shares their name with a famous rapper.
a. Eminem   b. Drake   c. Tupac   d. Nelly

35. What is the name of the famous football stadium in Rio de Janeiro that can hold up to 95,000 people?

36. Which country is at the northernmost point of South America?

37. Which country is at the westernmost point of South America?

38. What was Pablo Escobar's nationality?

39. As of September 2021, what is the total population of South America?
a. 378,845,235 b. 435,122,271
c. 663,792,554 d. 792.100,327

40. What is the name of the statue of Christ with his arms outstretched in Rio?

OCEANIA

1. Which country's capital is Port Vila?

2. In which Australian state would you find the city of Brisbane?

3. Name the only major city in Western Australia.

4. What is the most populous city in Oceania?

5. Name the indigenous population of New Zealand.

6. What is the capital of Fiji?

7. What is the world's largest coral reef system?

8. Name the island nation found in the Pacific Ocean which is the only country in the world to be in all four hemispheres.
a. Nauru   b. Kiribati   c. Solomon Islands

9. Name the subregion composed of volcanic islands stretching between Easter Island, New Zealand and Hawaii.

10. Name the subregion of Oceania which contains the Marshall Islands and the U.S Territory of Guam.
a. Micronesia   b. Palau   c. Mariana Islands

11. Name the subregion which covers Vanuatu, Tonga and Fiji.

12. Name the Islands which lie east of Papua New Guinea and is currently a Protectorate of the United Kingdom.

13. The Kingdom of _____ is a country and archipelago consisting of 169 islands.

14. The Independent State of _____ (which was previously known as Western _____ until 1997) is an island country found halfway between Hawaii and New Zealand.

15. The Republic of _____ is an island country found in the west Pacific. Its most populous island is Koror.

16. Which Strait separates the North and South Islands of New Zealand?

17. The majority of New Zealand's population live in which region?

18. Which island country was previously named 'Pleasant Island'?

19. Complete the name of the French territory located in the South Pacific: Wallis and _____.

20. What is the largest urban area in the South Island of New Zealand?
a. Auckland   b. Christchurch   c. Wellington

21. According to the United Nations, how many countries make up Oceania?
a. 10   b. 12   c. 14   d. 16

22. Approximately how many islands are there in Oceania?
a. 4,000 b. 10,000 c. 18,000 d. 27,000

23. What is the name of the supposed eighth continent in the world which contains New Zealand?

24. Oceania has traditionally been split into four different parts. These are Australasia, Micronesia, Polynesia and _____.

25. How many time zones does Australia have?
a. 1   b. 2   c. 3   d. 4

26. How many states does Australia have?
a. 4   b. 6   c. 8   d. 10

27. Which latitude line runs through Australia?
a. Tropic of Cancer   b. Tropic of Capricorn   c. Equator

28. Which state is the largest in Australia?

29. Approximately how long is the Great Barrier Reef?
a. 1000 km   b. 2000 km   c. 3000 km

30. How many states does New Zealand have?
a. 6   b. 9   c. 12   d. 16

ANTARCTICA

1. Is Antarctica at the North or South Pole?

2. True or false: Polar bears live in Antarctica.

3. What are the native people of Antarctica called?

4. True or false: Antarctica has the world's largest desert.

5. Not surprisingly Antarctica is the world's coldest continent. It is also the:
a. World's foggiest continent
b. Continent with the most earthquakes
c. World's windiest continent

6. True or false: Antarctica was once a warm and tropical continent.

7. When did humans first see Antarctica?
a. 1295 b. 1480 c. 1675 d. 1820

8. Around the summer solstice, how many hours of daylight does Antarctica have?
a. 0 b. 1 c. 23 d. 24

9. What is the name of the ship that became stuck in the ice during Shackelton's famous expedition across Antarctica in 1914?

10. Antarctica was part of which super continent 180 million years ago?
a. Pangea b. Laurasia c. Godwana

11. What is the average temperature at the South Pole?
a. -18 degrees   b. -31 degrees   c. -48 degrees

12. Listed in order of largest to smallest, where does Antarctica rank in terms of size compared to the other continents?
a. 2nd   b. 3rd   c. 4th   d. 5th

13. Approximately how much of Antarctica is covered by ice?
a. 84%   b. 89%   c. 93%   d. 98%

14. How many species of plants have been recorded in Antarctica?
a. 0   b. 2   c. 50   d. 350

15. From which language does Antarctica get its name?
a. Latin   b. Greek   c. Hebrew

16. If all the ice in Antarctica melted, how much would global sea levels rise by?
a. 45-50 metres   b. 60-65 metres
c. 80-85 metres   d. 115-120 metres

17. True or false: whaling is still allowed in Antarctica.

18. How many countries are in Antarctica?
a. 0   b. 1   c. 3   d. 6

19. What is the name of the highest volcano in Antarctica?
a. Mount Sidley   b. Mount Melbourne
c. Mount Terror   d. Mount Erebus

20. The world's largest glacier can be found in Antarctica, but what is its name?
a. Amundsen   b. Lamdert   c. Paramon

CAPITAL CITIES

1. What is the northernmost capital city in Europe?

2. What is Europe's southernmost capital city?

3. Which is the most populous capital city in Europe?
a. Paris   b. London   c. Athens   d. Moscow

4. Which South Pacific Island is widely regarded to be the only country in the world to not have a capital?
a. Kiribati   b. Nauru   c. Vanuatu   d. Fiji

5. What is the last world capital in alphabetical order?

6. What is the first world capital in alphabetical order?

7. What is the capital of Canada?

8. Munich is the capital of which German state?

9. Name the most southerly capital city in the world.

10. What is the largest capital city in the world by population?
a. New Delhi   b. Tokyo   c. Beijing

11. What is the largest capital city by area?
a. Beijing   b. Tokyo   c. Moscow   d. Jakarta

12. Which capital city is divided by the river Danube?

13. Which Middle Eastern capital city is considered the oldest in the world?
a. Damascus   b. Muscat   c. Kuwait City

14. In which capital city does the Hope River flow through?

15. Name the capital of the U.S state of Georgia.

16. What is the only capital city to begin with 'e'?

17. What word is added to the self-named capital city of Mexico, Guatemala, Kuwait, Luxembourg and Djibouti?

18. Following on from question 17, which is the only other country with the same word added to its capital?

19. What is the capital of Chile?

20. What is the capital of Morocco?

21. Which of the following capitals is the most northerly?
a. Stockholm   b. Reykjavik   c. Copenhagen

22. Which of these has never been an official capital of the United States?
a. Boston   b. New York   c. Philadelphia

23. Which of these capitals has the highest population?
a. Tokyo   b. Seoul   c. London   d. Moscow

24. What is the capital of Ethiopia?

25. In which of these capitals would many of the locals speak French?
a. Brussels   b. Vienna   c. Bern

26. What is the capital of Bulgaria?

27. In which capital city would you find the Little Mermaid statue?

28. In which continent would you find the city of Baku?

29. In which capital would you find the Forbidden City?

30. What is the capital of Nepal?

31. In which European capital would you find the Trevi Fountain?

32. WAW is the airport code for which capital city?

33. What is the capital of Venezuela?

34. Which of the following is the capital of Costa Rica?
a. San Cristobel   b. San Jose   c. San Maria

35. What is the capital city with the highest altitude in the world?
a. Quito   b. La Paz   c. Kathmandu

36. Vaduz is the capital of which country?

37. What is the capital of South Korea?

38. Four of the following eight countries have capital cities that
are the same name as the country. How many can you identify?
a. Djibouti   b. Panama   c. Kiribati
d. Fiji   e. Nauru   f. Singapore
g. Guyana   h. Sao Tome and Principe's

39. What is the capital of Somalia?

40. What is the capital of the Republic of Ireland?

HISTORICAL GEOGRAPHY

1. By which name was Thailand known as before 1939?

2. Which country was formerly known as Burma?

3. Name the supercontinent from the late Palaeozoic era which incorporated all the Earth's landmasses.

4. What is the previous name of Mumbai?

5. Which Asian capital was previously known as Edo?

6. Which Indian state was previously the capital of Portuguese India?

7. What was the capital of West Germany?

8. Which U.S state was named after a French king and was previously known as 'New France'?

9. The ancient city of Babylon would be found in which modern country?
a. Iraq   b. Afghanistan   c. Iran   d. Pakistan

10. Which West African nation was previously known as Gold Coast?
a. Mali   b. Mauritania   c. Ghana

11. Which African country was once known as Abssyinia?
a. Mali   b. South Africa   c. Ethiopia

12. By which name was Ho Chi Minh City in Vietnam previously known as?

13. What is East Pakistan known as today?

14. Which country was previously known as Ceylon?
a. Sri Lanka   b. Mongolia   c. Nepal

15. Persia refers to which modern day country?

16. Which capital city was previously known as Byzantium, Constantinople and New Rome?

17. Which major Asian city was previously known as Peking?

18. Which South Pacific Island state was previously known as Van Diemen's Land?

19. Which European capital was previously known as Kristiania?
a. Helsinki b. Oslo c. Copenhagen d. Sofia

20. As of 2018, what is the new name for the Kingdom of Swaziland?

OCEANS AND SEAS

1. Name the European country which is surrounded by four seas.

2. Which is the only sea without any coasts?
a. Caspian Sea   b. Sargasso Sea
c. Laptev Sea   d. Baltic Sea

3. Which country has the longest coastline in the world?

4. What is the deepest point in the Earth's oceans called?

5. In which ocean can 75% of the Earth's volcanoes be found?

6. Which is the coldest sea in the world?
a. The White Sea   b. The Red Sea
c. The Black Sea   d. The Yellow Sea

7. The Seychelles are in which ocean?

8. Which is the only country to border both the Caspian Sea and the Persian Gulf?
a. Pakistan   b. Iraq   c. Iran   d. Yemen

9. Which sea is bordered by Queensland, Australia?

10. Which body of water is closest to Antarctica?

11. Which body of water separates North America from Asia?

12. What is the colour of Senegal's Lake Retba? It is also known as Lac Rose.
a. Red   b. Pink   c. Orange   d. Blue

13. Name the sea in which the islands Hvar, Vis and Mijet are found.
a. Mediterranean Sea   b. Baltic Sea
c. Black Sea   d. Adriatic Sea

14. In which ocean would one find the answers to question four?

15. Which sea separates Europe and Africa?

16. Which gulf lies along the western coast of France and the northern coast of Spain?
a. Bay of Biscay   b. Bay of Kotor
c. Bay of Plenty   d. Bay of Tonkin

17. Name the deep-water gulf found between Yemen, Djibouti and Somalia. The Gulf of ____. (4 letters long)

18. How many oceans does the USA border?

19. What is the name of the warm ocean current that runs through the Atlantic Ocean?

20. How much of the Earth's surface is covered by oceans?
a. 62%   b. 66%   c. 71%   d. 76%

LANGUAGE

1. Which European language is an official language of Macau, alongside Cantonese?

2. Khmer is spoken by more than 95% of which Asian country? (It is also the world's largest alphabet, with 74 letters in it)
a. Mongolia   b. Turkmenistan   c. Cambodia

3. Which Asian country's name means 'Land of the Pure' in Urdu and Persian?
a. Pakistan   b. Nepal   c. Bhutan   d. Oman

4. What is the most common first language in South Africa?

5. Which language has the second largest number of native speakers in the world?
a. English   b. Mandarin   c. Spanish

6. Which European language is the official language of Suriname?

7. What is the most widely spoken language in Europe?

8. Name the language which is considered a Low Franconian dialect and is spoken by 55% of Belgian citizens.

9. After English, what is the second most predominant language of the Caribbean?

10. What is the official language of Iran?

11. What is the official language of Pakistan?

12. Quechua is spoken by which indigenous group from the Americas?

13. What type of alphabet does Russian use?

14. Put the following 4 languages in order of how many native speakers they have in Europe: English, Turkish, French, Italian.

15. Which of the following is not a Germanic language: French, Dutch, Swedish?

16. How many of the following languages don't have alphabets: Korean, Japanese, Mandarin?

17. To the nearest 1000, approximately how many languages are spoken in the world today? A. 5000 b. 7000 c. 10000 d. 14000

18. Which continent has the most languages?

19. Most historians believe that the writing form of languages originated from which civilisation?
A. Mesopotamia   b. Ancient Egypt   c. Romans   d. Aztecs

20. What is the second most spoken language in Canada?

RIVERS AND LAKES

1. What is Europe's most used commercial waterway, which begins in the Black Forest of Germany and flows into the Black Sea?

2. What is the longest river in Europe?
a. Danube   b. Ural   c. Volga   d. Kama

3. Which river runs through Paris?
a. Rhine   b. Seine   c. Garonne

4. Which river is the widest in the world?

5. Name the river which flows through Cologne and Basel and empties into the North Sea.

6. This river flows through Germany, Poland and Czechia. It is also the German word for 'or'.

7. This is a famous Slovenian Lake only 35km from Ljubljana, which is a popular tourist destination.

8. This is an African lake which is also the name of an North-Central African country. From 1963-1998 it is thought to have shrunk up to 95%, and it provides water to Cameroon, Chad, Niger and Nigeria. (Hint one of the countries that it provides water to is the answer)

9. What is the largest freshwater lake in the world?
a. Lake Victoria   b. Lake Michigan
c. Lake Tanganyika   d. Lake Superior

10. What is the longest river in Asia?
a. Yangtze   b. Yellow River   c. Lena River

11. Which lake is the smallest of the North American Great Lakes?
a. Lake Michigan   b. Lake Huron
c. Lake Ontario   d. Lake Erie

12. This river is shared by Argentina, Paraguay and Brazil.
a. Orinoco   b. Rio de la Plata   c. Parana

13. Where would you find the Irrawaddy River?
A. Cambodia   b. Thailand   c. Myanmar

14. How long is the river Nile?
a. 6650 km   b. 8700 km   c. 10150 km

15. Which three countries does the Tigris River run through?
a. Iran   b. Iraq   c. Oman   d. Greece   e. Turkey   f. Syria

16. Australia's longest river shares its name with a tennis legend. What is the name of this river?

17. How many US States does the Mississippi flow through or border?
a. 8   b. 10   c. 12   d. 14

18. How many countries does the Danube run through?
a. 4   b. 6   c. 8   d. 10

19. What is the UK's longest river?
a. Thames   b. Severn   c. Trent   d. Wye

20. Which two European countries does the river Elbe run through?
a. Austria   b. Hungary   c. Germany   d. Czech Republic

FLAGS

1. Which is the only flag that is neither a rectangle nor a square?

2. Which of the following flags has a dragon on it?
a. Mongolia   b. Ethiopia   c. Micronesia   d. Bhutan

3. If you were to turn the Indonesian flag upside down, which European countries flag would you get?

4. When was the original stars and stripes flag designed for the USA?
a. 1589   b. 1643   c. 1726   d. 1777

5. What do the flags of Saudi Arabia, Mozambique, Guatemala and Belize have in common?
a. All feature a sunset   b. All feature animals
c. All feature weapons

6. Which country holds the Guiness World Record for the oldest continuously used national flag?
a. Denmark   b. Norway   c. India   d. Peru

7. Which three colours appear on the flag of Chile?

8. From 1815-1830, the flag of this country was just a white flag. Which country is this?
a. Japan   b. Iraq   c. Portugal   d. France

9. How many stars are on the Chinese flag?
a. 3   b. 4   c. 5   d. 6

10. Afghanistan holds the record for changing its flag the most of all countries. How many times has it changed its flag?
a. 12   b. 18   c. 25   d. 36

11. Which of the following flags features a sunrise?
a. Togo   b. Trinidad and Tobago
c. Suriname   d. Antigua and Barbuda

12. How many stars appear on the Australian flag?
a. 4   b. 5   c. 6   d. 10

13. Which of the following flags does not have the Union Jack on it?
a. Falkland Islands   b. Tuvalu   c. Fiji   d. Solomon Islands

14. How many red and white stripes appear on the US flag?
a. 7   b. 9   c. 11   d. 13

15. What is the study of flags called?

16. True or false: American law states that if the flag is no longer a fitting emblem for display, then it should be burnt.

17. What is special about the flags of Switzerland and the Vatican City?

18. What type of animal is there on the Colombian flag?

19. What colours appear on the flag of Finland?

20. How many stars appear on the Ghanaian flag?
a. 1   b. 2   c. 5   d. 8

NATURAL DISASTERS

1. In 1923, there was a massive earthquake that struck Tokyo, killing 99,000 people. The earthquake however was not responsible for killing most of the people. What was responsible?
a. Tsunami   b. Firestorms   c. Nuclear waste

2. The explosion of which volcano in 1669 caused a local war?

3. Where did the worst ever earthquake in terms of human life lost occur, where 830,000 lost their lives in 1556?
a. India   b. Chile   c. Mexico   d. China

4. The largest known tsunami occurred in 1971. Where did this monster wave occur?

5. In which year did the island of Krakatoa explode?
a. 1762   b. 1798   c. 1837   d. 1883

6. What is the name of the earthquake that erupted in Iceland in 1783, that killed 20% of the islands inhabitants and opened an 18-mile-long fissure?
a. Laki   b. Nori   c. Valdivia   d. Tohoku

7. Which of these scales is used to categorise hurricanes by their wind speed?
A. Richter scale   b. Saffir-Simpson scale   c. Fujita scale

8. What type of natural disaster wiped out entire villages in Iran in 1972?
A. Blizzard   b. Earthquake   c. Wildfire   d. Tsunami

9. In 1976, a 7.8 magnitude earthquake struck, killing over 240,000 people. In which country did this disaster occur?

10. The highest magnitude earthquake occurred on the 22nd of May in 1960, but which country was unfortunate enough to house this?

11. In which country did one of the worst typhoons ever occur in 1959, where approximately 5000 people lost their lives?
a. Thailand   b. Malaysia   c. Japan   d. North Korea

12. What caused the Pan American Airlines aircraft to crash on the 8th of December 1963?

13. Cyclone Tracy almost destroyed which Australian city on Christmas day in 1974?
a. Adelaide   b. Canberra   c. Darwin

14. Which of the following natural disasters can produce the highest winds?
A. Cyclone   b. Tornado   c. Hurricane

15. What are the odds of being struck by lightning in your lifetime?
a. 1 in 15,000   b. 1 in 50,000   c. 1 in 110,000

16. Which American city suffered from a very deadly earthquake in 1906?
a. Boston   b. Los Angeles   c. San Francisco

17. What is the name of the disease that is estimated to have killed over 50 million people in the early 20th century?

18. What type of cloud can form a tornado?
a. Stratus   b. Cumulonimbus   c. Cumulus

19. The distress call 'mayday' is globally used in radio communications to signal a life-threatening emergency. From which language does this word come from?

20. In which year did the great fire of London occur?
a. 1444   b. 1555   c. 1666   d. 1777

MOUNTAINS

1. What is the second highest mountain in the Alps?
a. Dom   b. Matterhorn   c. Monte Rosa

2. Name the mountain range located between Spain and France.

3. What is the oldest active volcano in the world?
a. Mauna Kea   b. Mount Etna
c. Mount Kilimanjaro   d. Mount Sidley

4. Which mountain range is found in Missouri and Arkansas?
a. Great Smoky Mountains   b. Rocky Mountains   c. Teton Range

5. Which mountain is the highest in Japan? Fuji

6. Which mountains form the boundary between Europe and Asia?

7. What is the highest mountain in New Zealand?
a. Mount Cook   b. Mount Tasman
c. North King   d. Mount Sefton

8. Where would you find the world's highest active volcano Mauna Loa?

9. What is the name of the Sherpa that assisted Hillary to the summit of Everest in 1953?

10. The following are the seven peaks: Denali, Kilimanjaro, Mount Elbrus, Mount Aconcagua, Everest, Vinson Massif, Mount Kosciuszko. Put them in order of their height.

11. What is the second highest mountain on earth?

12. What is the name of the highest peak in Great Britain?
a. Snowdon   b. Ben Nevis   c. Scafell Pike

13. Which of the following is the third highest mountain in the world?
a. Makalu   b. Lhotse   c. Kangchenjunga

14. There are 3 north faces in the Alps, known as the trilogy, that are famous for being some of the hardest to climb in the world. How many of these can you name?

15. The Himalayas pass through 6 different nations. How many of them can you name?

16. Which mountain did Alex Honnold famously 'free solo' (climbing without ropes) in 2017?

17. What is the name of the Iceman that was found in Oetztal in the Alps in 1991?

18. What is the name of the famous New Zealand Mountain guide that lost his life in 1996 whilst leading his clients to the summit of Everest? As he sat in the snow succumbing to hypothermia, he used the last of his radio battery to ring his pregnant wife, and was never heard from again.

19. The 'death zone' refers to a certain height that climbers reach when climbing the highest mountains in the world, where there is so little oxygen which has resulted in hundreds of people dying of altitude sickness. What height is known as the 'death zone'?
A. 7000m b. 7500m c. 8000m d. 8500m

20. What is the second highest mountain in the United States?
a. Mount Foraker   b. Mount Saint Elias
c. Mount Bona   d. Mount Blackburn

GENERAL KNOWLEDGE

1. Which country has the tallest building in Europe?
a. England   b. France   c. Russia   d. Italy

2. Name the largest landlocked country in the world.
a. Afghanistan   b. Turkmenistan
c. Kazakhstan   d. Azerbaijan

3. Which country is the highest in the world, with an average elevation of 10, 761 feet?
a. Nepal   b. Bhutan   c. Pakistan

4. The capital city of which country is widely believed to be the oldest city in the world?
a. Greece   b. Italy   c. Syria   d. Australia

5. What is the smallest country in the world by size?

6. Which continent is located within all four hemispheres?

7. Which continent contains the most fresh water?

8. Asia is the world's largest continent by size. Which is the second largest?

9. The Galápagos Islands are part of which country?
a. Colombia   b. Paraguay   c. Ecuador

10. What is the closest non-bordering country to the USA?

11. Name the country which is the only one in the world to begin with this vowel.

12. What is the only communist country in the western hemisphere?

13. Which of the '-stan' countries is the most populous?

14. Greenland and the Faroe Islands are overseas dependencies of which country?
a. Norway   b. Sweden   c. Denmark

15. What is the largest Hindu country in regard to population percentage?
a. Bangladesh   b. India   c. Nepal

16. If all the countries in the world were listed alphabetically, which country would come second?

17. Name the cheese which is named after a Dutch town of the same name.

18. Which continent is the youngest in the world, with over 60% of its population under the age of 25?

19. Which continent (not including Antarctica) is the driest?

20. Which major city is in two continents?
a. Dhaka   b. Mexico City   c. Istanbul

21. Which nation produces two thirds of the world's vanilla?
a. Somalia   b. Burkina Faso   c. Madagascar

22. A sabra is a native person from which country?

23. As of 2020, what is the newest country in the world?

24. Name the Indian public holiday which is also known as the Hindu festival of lights.

25. Which country has 62% of the world's lakes?

26. Name the island where you will find the world's most northerly community.

27. Which country has the highest GDP per capita in the world?
a. Qatar   b. Luxembourg
c. Monaco   d. Norway

28. In which country would you find the Dead Sea?

29. Name the region which covers around 75% of Russia.

30. Name an indigenous group from Alaska.

31. Name the world's oldest country, where over a quarter of its population is over the age of 65.

32. Name the world's youngest country, with over 50% of its population under the age of 15.
a. Myanmar b. Tonga c. Somalia d. Niger

33. In which European country is there no airport and no trains?
a. Luxembourg b. Vatican City c. Andorra

34. Where would you find the world's largest national park?

35. How many time zones does Russia have?
a. 4 b. 6 c. 8 d. 11

36. Name the Islands which were named after a Latin phrase meaning 'island of dogs'.
a. Canary Islands   b. Cayman Islands
c. Cook Islands   d. Caribbean

37. Name the Spanish Island which is located off the coast of West Africa.

38. Which is the largest island in the Mediterranean Sea?

39. True or false: the Pacific Ocean is larger than the moon.

40. Which category do corals fall under?
a. Animals   b. Plants   c. Fungus

MATCH THE COUNTRY TO ITS HEMISPHERE

This round will give you the name of a country. You must decide whether it is in the Southern or Northern hemisphere.

1. Mexico

2. Bangladesh

3. Vietnam

4. Russia

5. Mauritania

6. Argentina

7. East Timor

8. North Korea

9. Botswana

10. Cambodia

11. Angola

12. Bhutan

13. Tuvalu

14. Micronesia

15. Costa Rica

16. Afghanistan

17. Oman

18. Egypt

19. Indonesia

20. Tonga

TOURISM

1. Which famous Paris landmark is the world's most visited art museum?

2. Name the Italian resort island which can be found just south of Naples.

3. Where would you find the Petronas Towers?
a. Singapore   b. Cambodia   c. Malaysia

4. Name South Africa's most famous National Park.

5. Name the ancient Inca city which can be found in Peru.

6. Name Iceland's natural, geothermal spa which attracts over a million visitors a year.

7. Name the famous prehistoric monument found in Wiltshire, England.

8. Name the ancient ruins which overlook Greek's capital.

9. Name the popular tourist destination which is found in between New York and Ontario.

10. Which country has the most UNESCO World Heritage Sites?
a. France   b. Spain   c. United Kingdom   d. Italy

11. Name the Islands where you would find Majorca.

12. In which country would you find the original Lego-Land?
A. Norway   b. Denmark   c. France   d. Germany

13. In which country was the Statue of Liberty constructed?

14. What colour were the pyramids at Giza originally?

15. In which country would you find Angkor Wat?

16. In which country would you find Petra?

17. What is the name of the Buddhist temple in Indonesia that was built in the 9th century, and is the top tourist attraction in Indonesia?

18. In which country would you find Bagan?
a. Myanmar   b. Mauritania   c. Mauritius

19. When did the Sydney Opera House officially open?
a. 1957   b. 1965   c. 1973   d. 1980

20. What is the name of the tallest building in the world (as of September 2021), that stands at 828m high and can be found in Dubai?

EXTRA HARD

1. What is the previous name of Kazakhstan's capital?

2. What is the currency of Kyrgyzstan?

3. What is the largest region of Peru?

4. Name the river which starts in Colorado, USA and drains in the Gulf of Mexico. It forms the border between Texas and Mexico.

5. Name the capital of Tuvalu.

6. What is the capital of Papua New Guinea?

7. What is the largest city in the Philippines by area?
a. Davao City b. Zamboanga c. Butuan City

8. Name the desert between the Nile and the Red Sea.

9. Name the most populated island in the world. It has a population of 136 million, and a living area of only 127,569 km^2.

10. Name the Caribbean Island which is home to two different countries.

11. What is the southernmost province of China?

12. Name the peninsula in Northern Europe which forms mainland Denmark, northern Germany and separates the North and Baltic Sea.

13. Which currency was used in Greece before the adoption of the Euro in 2001?

14. Name the ancient trade route which connected the West with the Middle East and the rest of Asia.

15. What is the name for the line of longitude which is defined as 0 degrees?

16. Paraguay is divided into two provinces: the Occidental and the
_____.

17. Roughly, what is the equatorial circumference of Earth?
a. 20,000 km   b. 30,000 km   c. 40,000 km

18. Brazil shares a border with every South American country bar two. Which two are they?

19. What is the name of the world deepest lake and where can it be found?

20. Which country's territory is geographically closest to the North Pole?
a. Canada   b. Sweden   c. Norway   d. Iceland

TOP 10'S ROUND

This round will ask you to name the top 10 things for various geography related topics.

The scoring for this round will be slightly different. You will get five points if your answer is in the correct position, three points if it is one position away, and one point if it is two positions away. For example, if you put China first and it was in fact first then this would give you five points, if you put Colombia first but it was actually third then you would get one point.

TOP 10 COUNTRIES BY POPULATION

1.

2.

3.

4.

5.

6.

7.

8.

9.

10.

As of September 2021

TOP 10 COUNTRIES WITH THE HIGHEST GDP PER CAPITA

According to worldpopulationreview.com

1.

2.

3.

4.

5.

6.

7.

8.

9.

10.

As of September 2021

FIRST 10 COUNTRIES IN ALPHABETICAL ORDER

1.

2.

3.

4.

5.

6.

7.

8.

9.

10.

LAST 10 COUNTRIES IN ALPHABETICAL ORDER

1.

2.

3.

4.

5.

6.

7.

8.

9.

10.

TOP 10 LARGEST ECONOMIES IN THE WORLD

1.

2.

3.

4.

5.

6.

7.

8.

9.

10.

TOP 10 LARGEST CITIES IN THE WORLD

According to archdaily.com

1.

2.

3.

4.

5.

6.

7.

8.

9.

10.

TOP 10 LARGEST STATES IN AMERICA BY AREA

1.

2.

3.

4.

5.

6.

7.

8.

9.

10.

FIRST 10 US STATES IN ALPHABETICAL ORDER

1.

2.

3.

4.

5.

6.

7.

8.

9.

10.

LAST 10 US STATES IN ALPHABETICAL ORDER

1.

2.

3.

4.

5.

6.

7.

8.

9.

10.

TOP 10 LARGEST CITIES IN THE US

1.

2.

3.

4.

5.

6.

7.

8.

9.

10.

TOP 10 LARGEST COUNTRIES IN EUROPE BY POPULATION
(INCLUDING RUSSIA)

1.

2.

3.

4.

5.

6.

7.

8.

9.

10.

TOP 10 LARGEST COUNTRIES IN AFRICA BY POPULATION

1.

2.

3.

4.

5.

6.

7.

8.

9.

10.

TOP 10 LARGEST COUNTRIES IN SOUTH AMERICA BY POPULATION

1.

2.

3.

4.

5.

6.

7.

8.

9.

10.

TOP 10 LARGEST COUNTRIES IN ASIA BY POPULATION

1.

2.

3.

4.

5.

6.

7.

8.

9.

10.

ANAGRAM ROUND

The following are anagrams of 20 different countries around the world. How many can you get?

1. agnot

2. big mule

3. adanac

4. aorta fichus

5. aero cud

6. acts cairo

7. moan

8. tyre uk

9. and clots

10. aland wezen

11. au gu ruy

12. arther nook

13. ana pj

14. armada scag

15. alia mos

16. iijf

17. ay irs

18. enemy

19. aila yams

20. ar qat

ANAGRAM ROUND

The following are anagrams of 20 different capital cities around the world. How many can you get?

1. air co

2. tao taw

3. arel sig

4. arne crab

5. a nevin

6. ink ms

7. a purge

8. ezra gb

9. iq out

10. non old

11. il nebr

12. as uv

13. elhi sink

14. cra ca

15. ar then

16. art kaja

17. ak jerky vi

18. ok toy

19. ari boni

20. lett lava

ANAGRAM ROUND

The following are 20 different anagrams of 20 different oceans, seas, lakes and rivers. How many can you get? (Hint: try to look for ocean, lake, river or sea in the anagram)

1. adan coniine

2. ania spaces

3. line

4. arne abasia

5. accent cairo

6. aec capo ficin

7. alike pourers

8. alika evictor

9. inver zamora

10. crin groove

11. aile reek

12. ara close

13. andree reanimates

14. at gey zn

15. amer thrives

16. anacreon tactic

17. abdel elk

18. ans egg

19. henri

20. ade ser

GEOGRAPHY TERMINOLOGY

1. Which word describes two points that are on exact opposite sides of the world? For example, the North and South Pole.

2. This word describes a chain or group of closely scattered islands. They are usually formed by volcanoes.

3. What is the name of a small bay that is sheltered, and is usually shaped like a horseshoe?

4. This word describes an area of land that has been built up at the mouth of a river.

5. Which word describes a hill that is made of sand?

6. Which word describes the area where fresh water from a river meets saltwater from the ocean?

7. Which word describes a long narrow inlet in between steep cliffs?

8. This word describes a hot spring which will sometimes shoot up water and steam.

9. Which word describes an area of ocean that is partially surrounded by land?

10. Which word describes a wetland area near rivers or lakes?

11. Which word describes areas in deserts that have water and lots of vegetation?

12. This word describes an area of land that is surrounded by water on three sides.

13. This word describes a series of mountains.

14. Which word describes a small river that flows into a larger river?

15. This word describes a flat area with very little vegetation where the soil is frozen.

16. This word describes a ring of coral that makes an island.

17. Which word describes the part of the earth that is capable of supporting life?

18. Which word describes a deep valley with steep sides?

19. Which word describes a shallow body of water that has a sandbank or strip of land separating it from the ocean?

20. Which word describes a large growth of coral under the sea?

Answers

EUROPE

1. Monaco
2. San Marino
3. Romania
4. Andorra, Monaco and the Vatican City
5. London
6. Denmark
7. Norway
8. France
9. Glasgow
10. Isle of Wight
11. Portugal
12. Russia
13. Switzerland
14. Mount Elbrus - although Mont Blanc is often considered to be Europe's highest mountain, this title is held by this Russian mountain because it lies on the European continental plate.
15. Lira
16. Liechtenstein
17. Malta
18. Sicily
19. Spain
20. Belarus
21. Norway
22. Austria
23. Barcelona
24. 35,000
25. Italy
26. 75 metres
27. Croatia
28. England, Scotland and Wales
29. Toulouse
30. Moldova
31. Lake Ladoga
32. France
33. Germany
34. 44
35. Kosovo
36. 1993
37. 2016
38. 12
39. 748,186,266
40. Portugal

AFRICA

1. Sudan
2. Lesotho
3. Nigeria
4. Kilimanjaro
5. Somalia
6. Djibouti
7. Tunisia
8. Eritrea
9. Uganda
10. Algeria
11. Lagos
12. Sahara
13. Angola
14. Seychelles - it only has 92,000 inhabitants!
15. Gambia
16. Kingdoms of Swaziland, Lesotho and Morocco
17. Tanzania
18. Kenya
19. Guinea
20. Benin
21. Senegal
22. Zimbabwe
23. Namibia
24. Tanzania
25. Zimbabwe
26. Mali
27. Casablanca
28. Alexandria
29. Cape Town
30. Pacific Ocean
31. Zambezi River
32. Mossi
33. Suez Canal
34. Islam
35. Swahili
36. 7th
37. Prime meridian
38. True
39. 54
40. 3000

NORTH AMERICA

1. Alaska
2. Maine
3. Yellowstone
4. Texas
5. Puerto Rico
6. Barbuda
7. Ontario
8. Death Valley
9. Dominica
10. Barbados
11. Cuba
12. El Salvador
13. Belize
14. Nunavut
15. Michigan
16. Wyoming
17. Aruba
18. Nicaragua
19. Newfoundland and Labrador
20. Pacific, Atlantic and Arctic
21. 13
22. 6
23. Canada
24. Baja 1000
25. Nahuas
26. 31
27. Jaguar
28. Hawaii
29. South Dakota
30. Washington
31. Sweden
32. Ontario
33. 5
34. Victoria
35. 23
36. Haiti and Honduras
37. 700
38. True
39. Costa Rica
40. Nicaragua

78

ASIA

1. Saudi Arabia
2. Borneo
3. Jerusalem
4. Mongolia
5. Indonesia
6. Uzbekistan
7. Gobi Desert
8. Macau
9. Baht
10. Pakistan
11. Maldives
12. Brunei
13. Tajikistan
14. Thailand
15. Yemen
16. Azerbaijan
17. Bangladesh
18. Bali
19. Goa
20. Taiwan
21. 48
22. 59.8%
23. Mongolia
24. Ho Chi Minh City
25. Laos
26. Laos
27. Malaysia
28. Maldives
29. Dead Sea
30. Origami

31. Jaya Peak
32. Mecca
33. 20 years
34. Sri Lanka
35. Mekong River
36. Once every 4 years
37. 29
38. China
39. Bengali
40. 1190

SOUTH AMERICA

1. Brazil
2. Paraguay and Bolivia
3. Colombia
4. Andes
5. Aconcagua
6. Colombia
7. Chile
8. Peru
9. Patagonia
10. Angel Falls
11. Ecuador
12. Atacama
13. France
14. Portuguese
15. Falklands
16. São Paulo
17. Suriname
18. Georgetown
19. Bogota
20. Lake Titicaca
21. 9
22. 14
23. Chile
24. Peru
25. Ecuador
26. Bolivia
27. US dollar
28. French Guiana
29. Peru
30. Sao Paulo

31. False - it is the highest
32. A corn pancake
33. Colombia
34. Tupac
35. Maracanã stadium
36. Colombia
37. Peru
38. Colombian
39. 435,122,271
40. Christ the Redeemer

OCEANIA

1. Vanuatu
2. Queensland
3. Perth
4. Sydney
5. Maori
6. Suva
7. Great Barrier Reef
8. Kiribati
9. Polynesia
10. Micronesia
11. Melanesia
12. Solomon Islands
13. Tonga
14. Samoa
15. Palau
16. Cook
17. Auckland
18. Nauru
19. Wallis and Futuna
20. Christchurch
21. 14
22. 10,000
23. Zealandia
24. Melanesia
25. 3
26. 6
27. Tropic of Capricorn
28. Western Australia
29. 2000 km
30. 16

ANTARCTICA

1. South Pole
2. False
3. There are none
4. True
5. World's windiest continent
6. True
7. 1820
8. 24
9. Endurance
10. Godwana
11. -48 degrees
12. 5th
13. 98%
14. 350
15. Greek
16. 60-65 metres
17. False
18. 0
19. Mount Erebus
20. Lambert

CAPITAL CITIES

1. Reykjavik
2. Athens
3. Moscow
4. Nauru
5. Zagreb
6. Abu Dhabi
7. Ottawa
8. Bavaria
9. Wellington
10. Beijing
11. Tokyo
12. Budapest
13. Damascus
14. Kingston
15. Atlanta
16. Edinburgh
17. City
18. Panama
19. Santiago
20. Rabat
21. Reykjavik
22. Boston
23. Tokyo
24. Addis Ababa
25. Brussels
26. Sofia
27. Copenhagen
28. Asia - It's the capital of Azerbaijan
29. Beijing
30. Kathmandu
31. Rome
32. Warsaw
33. Caracas
34. San Jose
35. La Paz - this capital sits 3,869 meters above sea level.
36. Liechtenstein
37. Seoul
38. Djibouti's capital - Djibouti City, Panama's capital - Panama City, Singapore's capital - Singapore, Sao Tome and Principe's capital - Sao Tome.
39. Mogadishu
40. Dublin

HISTORICAL GEOGRAPHY

1. Siam
2. Myanmar
3. Pangea
4. Bombay
5. Tokyo
6. Goa
7. Bonn
8. Louisiana
9. Iraq
10. Ghana
11. Ethiopia
12. Saigon
13. Bangladesh
14. Sri Lanka
15. Iran
16. Istanbul
17. Beijing
18. Tasmania
19. Oslo
20. Eswatini

OCEANS AND SEAS

1. Italy (the Adriatic, Ionian, Mediterranean and Tyrrenhian)
2. Sargasso
3. Canada
4. Mariana Trench
5. Pacific Ocean
6. The White Sea
7. Indian Ocean
8. Iran
9. Coral Sea
10. Southern Ocean
11. Bering Straight
12. Pink
13. Adriatic Sea
14. Pacific Ocean
15. Mediterranean Sea
16. Bay of Biscay
17. Gulf of Aden
18. Three - The Atlantic, The Pacific and the Arctic
19. The Gulf Stream
20. 71%

LANGUAGE

1. Portuguese
2. Cambodia
3. Pakistan
4. Zulu
5. Spanish
6. Dutch
7. German
8. Flemish
9. French
10. Farsi
11. Urdu
12. Incas
13. Cyrillic
14. 1. French 2. Turkish 3. Italian 4. English
15. French
16. One - Chinese
17. 7000
18. Asia
19. Mesopotamia
20. French

RIVERS AND LAKES

1. Danube
2. Volga
3. Seine
4. Amazon
5. Rhine
6. Oder
7. Lake Bled
8. Chad
9. Lake Superior
10. Yangtze
11. Lake Ontario
12. Parana River
13. Myanmar
14. 6650 km
15. Iraq, Turkey and Syria
16. Murray river (Andy Murray)
17. 10
18. 10
19. Severn
20. Germany and Czech Republic

FLAGS

1. Nepal
2. Bhutan
3. Poland
4. 1777
5. All feature weapons
6. Denmark
7. Red, white and blue
8. France
9. 5
10. 25
11. Antigua and Barbuda
12. 6
13. Solomon Islands
14. 13
15. Vexillology
16. True
17. They are square
18. A bird - more specifically a condor
19. Blue and white
20. 1

NATURAL DISASTERS

1. Firestorms directly after the earthquake
2. Mount Etna
3. Shensi Province China
4. Philippines
5. 1883
6. Laki
7. Saffir-Simpson scale
8. Blizzard amazingly
9. China
10. Chile
11. Japan
12. Lightning
13. Darwin
14. Tornadoes
15. 1 in 15,000
16. San Francisco
17. Spanish flu
18. Cumulonimbus usually forms tornadoes
19. French - comes from the phrase m'aider
20. 1666

MOUNTAINS

1. Monte Rosa
2. Pyrenees
3. Mount Etna
4. Ozark
5. Fuji
6. Ural
7. Mount Cook
8. Hawaii
9. Tenzing Norgay
10. 1 - Everest   2 - Aconcagua
    3 - Denali   4 - Kilimanjaro
    5 - Mount Elbrus 6 - Vinson Massif
    7 - Kosciuszko
11. K2
12. Ben Nevis
13. Kangchenjunga
14. The Eiger, The Matterhorn and Grandes Jorasses
15. India, China, Nepal, Pakistan, Bhutan, Afghanistan
16. El Capitan
17. Ötzi the Iceman
18. Rob Hall
19. 8000 m
20. Mount Saint Elias

GENERAL KNOWLEDGE

1. Russia (Lakhta Centre)
2. Kazakhstan
3. Bhutan
4. Syria
5. Vatican City
6. Africa
7. Antarctica
8. Africa
9. Ecuador
10. Russia
11. Oman
12. Cuba
13. Pakistan
14. Denmark
15. Nepal
16. Albania
17. Edam
18. Africa
19. Oceania
20. Istanbul
21. Madagascar
22. Israel
23. South Sudan
24. Diwali
25. Canada
26. Svalbard
27. Luxembourg
28. Jordan
29. Siberia
30. Aleut, Yupik and Inuit
31. Japan
32. Niger
33. Andorra
34. Greenland
35. 11
36. Canary
37. Tenerife
38. Sicily
39. True
40. Animals

MATCH THE COUNTRY TO ITS HEMISPHERE

1. Northern hemisphere
2. Northern hemisphere
3. Northern hemisphere
4. Northern hemisphere
5. Northern hemisphere
6. Southern hemisphere
7. Southern hemisphere
8. Northern hemisphere
9. Southern hemisphere
10. Northern hemisphere
11. Southern hemisphere
12. Northern hemisphere
13. Southern hemisphere
14. Northern hemisphere
15. Northern hemisphere
16. Northern hemisphere
17. Northern hemisphere
18. Northern hemisphere
19. Southern hemisphere
20. Southern hemisphere

TOURISM

1. Louvre
2. Capri
3. Malaysia
4. Kruger
5. Machu Picchu
6. Blue Lagoon
7. Stonehenge
8. Acropolis
9. Niagara Falls
10. Italy
11. Balearic
12. Denmark
13. France
14. White
15. Cambodia
16. Jordan
17. Borobudur
18. Myanmar
19. 1973
20. Burj Khalifa

EXTRA HARD

1. Astana
2. Lom
3. Loreto
4. Rio Grande
5. Funafuti
6. Port Moresby
7. Davao
8. Nubian
9. Java
10. Hispaniola
11. Hainan
12. Jutland
13. Drachma
14. Silk Road
15. Prime Meridian
16. Oriental
17. 40,000 km
18. Chile and Ecuador
19. Baikal, Russia
20. Canada

TOP 10 COUNTRIES BY POPULATION

1. China — population of 1,439,323,776
2. India — population of 1,380,004,385
3. United States — population of 331,002,651
4. Indonesia — population of 273,523,615
5. Pakistan — population of 220,892,340
6. Brazil — population of 212,559,417
7. Nigeria — population of 206,139,589
8. Bangladesh — population of 164,689,383
9. Russia — population of 145,934,462
10. Mexico — population of 128,932,753

TOP 10 COUNTRIES BY HIGHEST GDP PER CAPITA

1. Luxembourg   GDP per capita: $118,001
2. Singapore   GDP per capita: $97,057
3. Ireland   GDP per capita: $94,392
4. Qatar   GDP per capita: $93,508
5. Switzerland   GDP per capita: $72,874
6. Norway   GDP per capita: $65,800
7. United States   GDP per capita: $63,416
8. Brunei   GDP per capita: $62,371
9. Hong Kong   GDP per capita: $59,520
10. Denmark   GDP per capita: $58,932

FIRST 10 COUNTRIES IN ALPHABETICAL ORDER

1. Afghanistan
2. Albania
3. Algeria
4. Andorra
5. Angola
6. Antigua and Barbuda
7. Argentina
8. Armenia
9. Australia
10. Austria

LAST 10 COUNTRIES IN ALPHABETICAL ORDER

1. Zimbabwe
2. Zambia
3. Yemen
4. Vietnam
5. Venezuela
6. Vanuatu
7. Uzbekistan
8. Uruguay
9. United States
10. United Kingdom

TOP 10 LARGEST ECONOMIES IN THE WORLD

1. United States   (GDP: 20.49 trillion)
2. China   (GDP: 13.4 trillion)
3. Japan: (GDP: 4.97 trillion)
4. Germany: (GDP: 4.00 trillion)
5. United Kingdom: (GDP: 2.83 trillion)
6. France: (GDP: 2.78 trillion)
7. India: (GDP: 2.72 trillion)
8. Italy: (GDP: 2.07 trillion)
9. Brazil: (GDP: 1.87 trillion)
10. Canada: (GDP: 1.71 trillion)

TOP 10 LARGEST CITIES IN THE WORLD

1. Tokyo   population of 37,339,804
2. New Delhi   population of 31,181,376
3. Shanghai   population of 27,795,702
4. Sao Paulo   population of 22,237,472
5. Mexico City population of 21,918,936
6. Dhaka   population of 21,741,090
7. Cairo   population of 21,322,750
8. Beijing   population of 20,896,820
9. Mumbai   population of 20,667,656
10. Osaka   population of 19,110,616

TOP 10 LARGEST STATES IN AMERICA

1. Alaska   area (square km) 1,700,139
2. Texas   area (square km) 695,676
3. California   area (square km) 424,002
4. Montana   area (square km) 380,850
5. New Mexico   area (square km) 314,939
6. Arizona   area (square km) 295,276
7. Nevada   area (square km) 286,368
8. Colorado   area (square km) 269,620
9. Oregon   area (square km) 254,819
10. Wyoming   area (square km) 253,349

FIRST 10 US STATES IN ALPHABETICAL ORDER

1. Alabama
2. Alaska
3. Arizona
4. Arkansas
5. California
6. Colorado
7. Connecticut
8. Delaware
9. Florida
10. Georgia

LAST 10 US STATES IN ALPHABETICAL ORDER

1. Wyoming
2. Wisconsin
3. West Virginia
4. Washington
5. Virginia
6. Vermont
7. Utah
8. Texas
9. Tennessee
10. South Dakota

TOP 10 LARGEST CITIES IN THE US

1. New York City   population of 8,622,357
2. Los Angeles   population of 4,085,014
3. Chicago   population of 2,670,406
4. Houston   population of 2,378,146
5. Phoenix   population of 1,743,469
6. Philadelphia   population of 1,590,402
7. San Antonio   population of 1,579,504
8. San Diego   population of 1,469,490
9. Dallas   population of 1,400,337
10. San Jose   population of 1,036,242

TOP 10 LARGEST COUNTRIES IN EUROPE BY POPULATION

1. Russia   population of 145,912,025
2. Germany   population of 83,900,473
3. United Kingdom   population of 68,207,116
4. France   population of 65,426,179
5. Italy   population of 60,367,477
6. Spain   population of 46,745,216
7. Ukraine   population of 43,466,819
8. Poland   population of 37,797,005
9. Romania   population of 19,127,774
10. Netherlands   population of 17,173,099

TOP 10 LARGEST COUNTRIES IN AFRICA BY POPULATION

1. Nigeria   population of 206,139,589
2. Ethiopia   population of 114,963,588
3. Egypt   population of 102,334,404
4. DR Congo   population of 89,561,403
5. South Africa population of 59,308,690
6. Tanzania   population of 59,734,218
7. Kenya   population of 53,771,296
8. Uganda   population of 45,741,007
9. Algeria   population of 43,851,044
10. Sudan   population of 43,849,260

TOP 10 LARGEST COUNTRIES IN SOUTH AMERICA BY POPULATION

1. Brazil   population of 212,559,417
2. Colombia   population of 50,882,891
3. Argentina   population of 45,195,774
4. Peru   population of 32,971,854
5. Venezuela   population of 28,435,940
6. Chile   population of 19,116,201
7. Ecuador   population of 17,643,054
8. Bolivia   population of 11,673,021
9. Paraguay   population of 7,132,538
10. Uruguay   population of 3,473,730

TOP 10 LARGEST COUNTRIES IN ASIA BY POPULATION

1. China   population of 1,439,323,776

2. India   population of 1,380,004,385

3. Indonesia   population of 273,523,615

4. Pakistan   population of 220,892,340

5. Bangladesh   population of 164,689,383

6. Japan   population of 126,476,461

7. Philippines   population of 109,581,078

8. Vietnam   population of 97,338,579

9. Turkey   population of 84,339,067

10. Iran   population of 83,992,949

ANAGRAM ROUND
COUNTRIES

1. Tonga
2. Belgium
3. Canada
4. South Africa
5. Ecuador
6. Costa Rica
7. Oman
8. Turkey
9. Scotland
10. New Zealand
11. Uruguay
12. North Korea
13. Japan
14. Madagascar
15. Somalia
16. Fiji
17. Syria
18. Yemen
19. Malaysia
20. Qatar

CAPITAL CITIES

1. Cairo
2. Ottawa
3. Algiers
4. Canberra
5. Vienna
6. Minsk
7. Prague
8. Zagreb
9. Quito
10. London
11. Berlin
12. Suva
13. Helsinki
14. Accra
15. Tehran
16. Jakarta
17. Reykjavik
18. Tokyo
19. Nairobi
20. Valletta

ANAGRAM ROUND

RIVERS, OCEANS, LAKES AND SEAS

1. Indian Ocean
2. Caspian Sea
3. Nile
4. Arabian Sea
5. Arctic Ocean
6. Pacific Ocean
7. Lake Superior
8. Lake Victoria
9. Amazon River
10. Congo River
11. Lake Erie
12. Coral Sea
13. Mediterranean Sea
14. Yangtze
15. River Thames
16. Antarctic Ocean
17. Lake Bled
18. Ganges
19. Rhine
20. Red Sea

GEOGRAPHY TERMINOLOGY

1. Antipodes
2. Archipelago
3. Cove
4. Delta
5. Dune
6. Estuary
7. Fjord
8. Geyser
9. Gulf
10. Marsh
11. Oasis
12. Peninsula
13. Range
14. Tributary
15. Tundra
16. Atoll
17. Biosphere
18. Canyon
19. Lagoon
20. Reef

BY
PARAGON PUBLISHING

If you have enjoyed this book, be sure to look at the many other quiz and crossword books that Paragon Publishing has to offer.

Printed in Great Britain
by Amazon